FIRE!
MY PARENTS' STORY

JESSIE HAAS

FIRE!
MY PARENTS' STORY

Greenwillow Books
New York

Printed in the United States of America

First Edition   10 9 8 7 6 5 4 3 2 1

Library of Congress Cataloging-in-Publication Data

Haas, Jessie.

  Fire! my parents' story / by Jessie Haas.

    p.   cm.

  Summary: Describes a nighttime fire in a country house in Vermont in 1948 and the struggle of the residents to save their belongings.

  ISBN 0-688-15203-1

  1. Fires—Vermont—East Dover—History—
20th century—Juvenile literature.

  2. East Dover (Vt.)—History—Juvenile literature.

  3. Haas family—Juvenile literature.

  4. Yeaw family—Juvenile literature.

  [1. Fires—Vermont.]   I. Title.

  F59.E15H33   1998   974.3'9—dc21

  97-11202   CIP   AC

For all who were there—
especially my grandmother,
Jessie Atherton Trevorrow

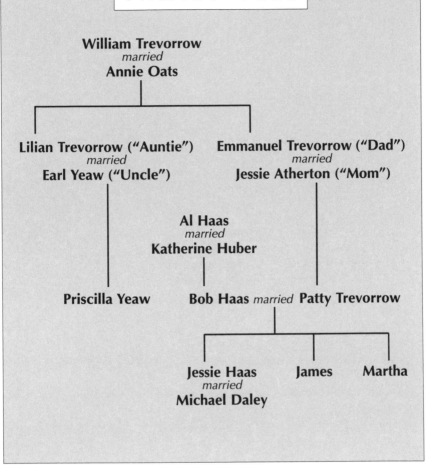

FAMILY TREE

William Trevorrow
married
Annie Oats

Lilian Trevorrow ("Auntie")
married
Earl Yeaw ("Uncle")

Emmanuel Trevorrow ("Dad")
married
Jessie Atherton ("Mom")

Al Haas
married
Katherine Huber

Priscilla Yeaw

Bob Haas *married* **Patty Trevorrow**

Jessie Haas
married
Michael Daley

James

Martha

chapter (*1*)

MAY 3, 1948, THREE O'CLOCK in the morning. High on a hill in East Dover, Vermont, where two dirt roads come together, an old white farmhouse dreams in the moonlight. Inside, people sleep, and across the road in the barns sleep horses, cattle, sheep, and pigs.

From the back wall of the kitchen, a tiny curl

of smoke is starting to rise. The smoke coils up the outside wall and in again through an open upstairs window.

Who is in the house?

In the shed the dog, Rover, is awake. He's worried, and whining.

Cousin Priscilla's bedroom is at the far end of the house, away from the smoke. Priscilla is fifteen, and she has to get up early tomorrow for the long drive to the Brattleboro high school. She sleeps soundly.

In the bedroom next to Priscilla's, Uncle Earl and Aunt Lilian sleep soundly, too. This is Uncle's house. He was born in it, and his father, Frank Yeaw, was born in it before him. Uncle is a carpenter. Auntie takes in boarders, city people who pay to stay here, breathe the country air and eat good food, walk on the dirt roads, hunt and fish and play at farming.

Al Haas is here now, all the way from Glendale, New York. He sleeps upstairs in the bedroom above Priscilla's. Al and his family have vacationed in this house for twenty years.

Al's son, Bobby, snores away in the little room at the head of the stairs. Bobby is eighteen. He graduated early from high school and was looking for a job when Uncle asked him to come help with sugaring. Sugaring is finished now. Bobby plans to work here on the farm until fall, when he'll go to college.

It's much quieter here than in Glendale. Al and Bobby sleep deeply.

The coil of smoke is growing. A fat column rises up the old white wall and rolls down the roof of the ell. The open window pulls it into the front bedroom. Inside, someone is coughing.

In the big bed near the window sleep Dad and Mom, Mandy and Jessie Trevorrow. They sleep hard because they have worked hard all day long. Dad runs Uncle's farm, and Mom works in the kitchen for Auntie, washing and cooking.

Dad is Auntie's brother. He came here as a little boy, after his mother died, and grew up on this farm. He met Mom when she was hired to work in the kitchen, and they fell in love.

Dad and Mom don't have a house of their own. They have the farm animals, all but the horses. Besides that they have only what is in this room: clothes, a hope chest filled with linen and china, a radio, a jar full of half-dollars, and an eight-year-old daughter, Patty.

The window is open because that's the way Dad likes to sleep, even when it's cold. Fresh air is good for you. There's a glass of water on the windowsill, where the outside air will keep it cool.

Smoke comes in the window now. It makes them sleep more deeply. If they breathe too much smoke, if they sleep too long, they might never wake up.

But in the other bed, away from the window, Patty coughs. She's still asleep, with her arm around her teddy bear.

Patty always has her arms around some animal: a cat, or Rover, or a lamb, or a cow, or, best of all, the workhorses, Chub and Sailor. Sailor is supposed to be a wild horse. Grown men are afraid of him. But both horses are Patty's friends. When she's been naughty, she hides in their stall and no one will come in after her. When she's good, she can drive them, all by herself.

The teddy bear is a pretend animal, but Patty hugs him because the cat, Tommy Tiger, is hiding under the bed away from the smoke.

The smoke hurts Patty's throat. She coughs harder and harder, until suddenly she coughs herself awake.

Her nose feels funny. Her eyes itch and her throat stings. She wants a drink of water.

"Mom?"

Mom doesn't answer.

"Mom? I need a drink of water.

"Mom?

"Mo-*om!*"

Slowly, slowly, Mom wakes up. "Hmm?" Her head aches. She can hardly think.

"Can I have some water?"

Mom gets out of bed and goes to the window. Fog is coming in, she thinks. It's been a wet, cool spring, with lots of fog.

But after a moment Mom smells smoke. Now she can see it, coiling down the roof, curling up over the sill into the room.

Quickly Mom turns toward Patty. She says, "Don't be afraid. The house is afire."

chapter (2)

PATTY SITS STRAIGHT up in bed. The house is afire? How can that be? There are no flames anywhere in sight. There is no red glow.

But Mom is shaking Dad's shoulder, and the bedsprings jounce. "Wake up!" she says. "*Fire!* Wake up!"

Patty's heart freezes. What's the matter? Why is Dad so slow? Why won't he wake?

But then he does. "Mmm?" he says. He rubs his hand over his head so his hair stands on end. He looks toward Mom.

"*Emmanuel!*" Mom says. "Get *up!* Patty, come with me!" She goes to the door and calls to Al and Bobby. "Wake up! Wake up! The house is afire!"

Patty gets out of bed, hugging her bear. The floor is cold under her bare feet. The house is dark and for a moment longer it is quiet. More quiet than before. Bobby has stopped snoring.

Dad just sits in bed. He says, "What?" His voice sounds fuzzy. Only Mom and Patty are up, waiting at the bedroom door.

Then—"*Hey!* Smoke!" Dad yells. His feet hit the floor. Al and Bobby come to the doors of their rooms.

"Where is it?"

"I don't hear anything!"

"Get out!"

"Wake the others!"

"Get Patty out!"

Patty clings to Mom. Mom is warm and firm, and Patty doesn't ever want to let go. The stairs look deep and black, like a cavern. What is down there? Where is the fire? There is no sound and the smoke is all up here, hanging in the air around their heads.

"Go on down!"

Mom tries to take a step, and Patty hangs on tighter. She's afraid to move. She wants to curl up tight. She wants this not to be happening. Mom can't pick her up. Mom has a very bad back.

"I'll take her," Bobby says. He picks Patty up under one arm. His pajama bottoms slip, and he grabs for them with the other hand. Down the black stairs he goes. Patty feels every step he takes, *bump bump bump*, like a scary carnival ride. She hangs on tight to her teddy bear. Mom and Dad and Al follow, and Tommy Tiger comes last of all.

At the bottom of the stairs Bobby turns and hurries through the dark dining room and the living room. He pushes the door open and rushes out onto the porch.

The air is cool and fresh. It smells like mud and new grass. There is no sign of a fire anywhere.

Bob sets Patty down and gives her a gentle push, down the stone step onto the cold, dewy lawn. Patty goes toward the road. She turns to look. Now she can see the thick snake of smoke coming up over the roof, swooping down, and then up and in through the bedroom window.

"It's out back!" Dad yells. He and Al and Bobby run around the corner.

But Mom doesn't come outside. She goes through the house to the other bedrooms. Patty can hear her trying to speak calmly.

"Now don't be afraid. The house is afire! Priscilla, wake up. The house is afire!"

"Where is it?" Uncle asks quickly.

Auntie says, "Oh no, oh no, oh no . . ." Then they're coming outside. Uncle and Priscilla run around back to see how big the fire is. Auntie comes down the lawn toward Patty, and then she stops, looking back at the house and the smoke. She squeezes her hands tight together. "Oh no."

But still Mom doesn't come out. She goes to the living room, to Uncle's old rolltop desk, and the telephone. She cranks the handle on the wall, making one long loud ring that tears through the dark house.

The long ring is the danger signal. It rings in all the houses on that party line. All the way down to East Dover village it startles people out of their sleep, it makes their hearts jump. It wakes up Mrs. Bush, the operator, and she picks up the receiver.

"This is Jessie up to Yeaws'," Mom says. "The house is afire."

Now Mom can hang up and run outside. Mrs. Bush knows just what to do.

First she calls the Brattleboro Fire Department. Then she tells all the people on the party line, who heard the long ring and stumbled to their telephones in the dark to pick up the receivers. They can all talk to one another at once, and Mrs. Bush can talk to them. "It's a fire up to

Yeaws'," she tells them, and they all hang up and run to get dressed.

Mrs. Bush calls other people now, on other lines. Out in the country like this, with no fire department close by, neighbors have to help one another. There is no one else.

chapter (3)

THE SUMMER KITCHEN, in the ell of the house, is big and airy. It has a wood stove and a new electric range, a broom closet, and a pantry. Mom and Auntie cook in the summer kitchen when it's hot and when there are lots of boarders. Now it is closed off to save heat.

But in the pantry the refrigerator is turned on. The fire is behind it. Already a hole has burned through the back corner of the house, in the exact shape of the refrigerator.

It's a small fire. If a fire truck came right now, and the firemen could turn on even their smallest hose, they would drown the flames in two minutes.

But the nearest fire truck is still twenty miles away. The alarm has just barely rung in the Brattleboro station. The firemen are just waking up. They haven't even reached for their clothes yet. They can't possibly get here in time to stop this fire while it's small.

Can Uncle and Dad put it out themselves? The hand pump is in the shed, next to the kitchen. It's connected to a good, free-flowing well. Two strokes of the handle will fill a pail with cold, pure water. Dad grabs a bucket and pushes the shed door open. Rover bursts out and races over to Patty.

Dad raises the pump handle for the first stroke. But something feels wrong. The handle is loose, and when he pushes it down it comes off in his hand. Yesterday the pump wasn't working right, and Uncle fixed it. He must not have put the handle together correctly.

For a second Dad stands looking at the handle. Loose in his hand, it is just a piece of metal. It can't bring the water up from underground. If there was time he could find the lost bolt, and thread a nut on it, and put the pump back together.

But in the next room the flames are eating into the wall with a fast, licking sound.

There's water in the bathroom and in the little winter kitchen. That water comes from a cistern in the attic. It flows down through small pipes, slowly.

Where else is water?

Across the road at the barns, in the tub where the animals drink! Dad starts to run. Bobby grabs a bucket and follows him. Rover follows, too. Rover doesn't like people to run or play roughly. Once he bit Bobby right in the rump, to stop him from wrestling with someone. But tonight Rover lets Dad and Bobby run.

They push open the big door and run through the horse barn, out into the muddy barnyard to the water tub. They plunge their buckets into the cold, black water. They haul them out again, heavy and dripping. They run back across the road as fast as they can, with the water sloshing on their legs, soaking their pajamas.

Already it's too late. The kitchen wall is old and dry. The fire gobbles it.

Dad and Bobby jump onto the back porch. The pantry is full of flames now, and the heat has broken the window. Dad and Bobby throw their buckets of water through the window. It doesn't make a bit of difference.

Uncle and Al are in the kitchen, trying to get the electric stove out. The door between the pantry and the kitchen is shut. No fire shows through, but the door is starting to smoke.

Dad goes to the window above the sink. He looks in at Uncle, and he just shakes his head.

Uncle comes out onto the front porch. He says to Auntie and Priscilla, to Mom and Patty, "Better start getting things out. It's going to go."

It's going to *go?* Uncle's voice has a sound that Patty has heard before. When something is completely broken. When an animal is too sick to live.

But the house is still here! Its white front gleams in the moonlight. The windowpanes twinkle. The roof comes down in its gentle, sheltering slope and the wide front porch is like a cave.

Why are they giving up? Patty wonders, out on the lawn hugging her bear. Couldn't they run for more water? Couldn't they *all* run?

But if that would work, they would be doing it. Dad can do anything in the world. Uncle and Auntie and Al and Bobby are smart people, and so is Cousin Priscilla, and Mom always knows what's best. They are not trying to save the house anymore. They are all rushing inside to save things.

chapter **(4)**

BOBBY RUNS UPSTAIRS for his pants and his wallet. He was paid yesterday. He has sixty dollars up there.

Priscilla rushes to her bedroom. She pushes the window open, and she starts throwing things out: her hairbrush, her books, her clothes. She is not crying, but her breath comes fast and catchy, like sobs.

Uncle goes back into the kitchen. He and Al wrestle with the big electric stove. They haul it and they rock it across the floor. It makes deep scratches in the wide boards, but that doesn't matter anymore. They slide the stove through the door onto the front porch. Bobby helps them get it down onto the ground.

Now Mom goes into the kitchen. Patty starts crying. She can't help it. She knows that's where the fire is.

But in a second Mom comes out. She brings the baby turkeys in their box, and then the ducklings in their washtub, and sets them on the porch.

"Take Patty somewhere safe!" Dad yells.

"Patty, come with me," Auntie says in a shaking voice. Auntie is terrified of fires. When she was a girl, her sister died in one. Auntie takes Patty's hand and they run to the car, parked beside the shed. Auntie drives the car across the road to the barn. "You'll be safe here," she says. She jumps out of the car again. "Stay right here," she tells Patty, and she runs back to the house. That is almost the strangest thing of all, seeing Auntie run.

The fire is getting bigger every second. It grows like fast vines along the shed-end, toward the road and the barns. Patty can see everyone hurrying in front of it, into the house, out again. She can't see Mom anywhere, but here comes Dad, running down the road.

He sees Patty in the car. "What are you doing here? It won't be safe!" Dad turns toward the house and yells. "Lilian! Lilian! Drive Patty up the hill!"

Patty doesn't know if anyone heard him. But Dad says, "I'll drive you myself, as soon as I let the stock out." He runs into the horse barn.

Chub and Sailor stand tied in their stalls. Chub nickers. Sailor takes three nervous steps—forward, back, to the side.

"Whoa," Dad says. He steps in quietly beside Chub. He unbuckles the halter and slips it off. Then he unbuckles Sailor's halter. The horses back out of their stalls and trot through the barn door.

Dad goes quickly into the dark, sweet-smelling cow barn. The cows all turn their heads. Their neck chains clink.

Dad hurries down the row of cows. Each neck chain has a different kind of snap or hook. Dad must unfasten each one by feel, reaching under the fat round necks of the Durhams and the narrow, elegant necks of the Guernseys, under the broad wet noses, careful of the horns. He releases each cow, and he heads them toward the door.

Now Dad runs to the little barn. He flings the sheep gate wide, and yells, and claps his hands. The sheep jump to their feet and crowd out past him, broad in their heavy wool coats. The lambs cling close to their sides.

Dad goes to the young cows. He unfastens their chains. The heifers jump back and blow their breath out loudly. Dad chases them outdoors.

But even in the barnyard, the animals are still too close to the burning house. It's just across the road, and the shed end is all ablaze. Dad can't see the fire, but he can see the big glow in the sky above the horse-barn roof.

Dad runs to the gate, but it is wired shut. This cold, rainy spring, none of the animals have been turned out to pasture yet. Until they are turned out, the gate is meant to stay closed, not to open easily.

Dad can't see the wire to unfasten it. Better

just tear the fence down instead. He takes hold of a fencepost and steps on the woven wire.

Something stabs, something feels like a knife, all across the bottoms of his feet. Patty hears him yell. Dad is barefoot. He didn't know it until just this minute.

He tramples the wire flat anyway. "Come on!" he calls. "Chub! Sailor!" The big horses gallop past him, out into the black pasture. "Kaday," he calls the sheep. "C'boss," he calls the cows.

The cows huddle at the other end of the barnyard. They won't come.

"Rover!" Dad calls. That scares the cows. They know if Rover comes he'll grab them by their tails, and sled along behind them on the ground. They mill around until they find the opening in the fence, and they rumble through. Rover streaks past the car and gets to the barnyard just as the sheep follow.

Patty hears the thunder of the animals galloping. She sees Dad run out of the barn with his hair all on end.

Then Auntie is opening the driver's side door. "I didn't think," she tells Patty, and she slides behind the wheel.

Auntie turns the car around fast, and the water in the ruts makes a ripping sound. She drives past the burning house, up the hill to the orchard. She parks the car beneath the big maple tree. It is much farther away than the barn.

"Now you stay right here in the car, no matter what," Auntie says. She jumps out and slams the door, and back she goes down the hill.

Patty tries to do what she's been told. She kneels on the broad, slippery seat. She presses her head against the cold windshield, and she looks down the hill.

But she can't see very well. She can't hear.

Orange reflections dance on the windshield. The long dark hood of the car seems to be in her way.

Patty opens the heavy door. She gets out of the car and stands in front of it, clutching her teddy bear hard around his middle.

She can see black figures moving in front of the fire. The ell is all ablaze now. The main part of the house is still all right, but flames are beginning to climb toward the bedroom windows.

The house doesn't notice. The windows gaze out calmly. The chimney stands still and straight, pointing toward the sky. The house can't feel the fire. It can't do anything. Each part looks the way it has always looked, until the flames reach it.

Everyone keeps going inside. But how fast will the fire spread? Will it catch them? Patty's tears come out hot. They cool off on their way down her face and drip onto the bear's head. Where is Mom?

Dad and Rover are running back from the barn, and everyone is yelling. Soon they are all yelling at Mom. "Jessie! Jessie, hurry up!"

Patty hugs her bear tighter, and she screams: "*Mom! Mom!*"

The old Yeaw house, 1947. The fire began in the summer kitchen, next to the shed at the right of this picture. The left-hand window above the porch was the one left open the night of the fire.

Bob Haas and Jessie Trevorrow
shell peas on the porch, 1947.
Patty Trevorrow peeks from
behind her mother.

Mandy Trevorrow plowing with Chub and Sailor. These horses were bred on the western range and brought to Brattleboro by train.

Mandy brings home a cow and new calf.

Al Haas, from a
German immigrant
neighborhood in
Glendale, New York,
vacationed on
the Yeaw farm
for over thirty years.

Rover chased cattle,
greeted guests, and
guarded Patty from
every possible threat.
This photo was the
farm Christmas card
for many years.

Patty in 1948, just weeks before the fire.

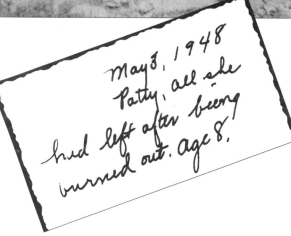

May 3, 1948
Patty, all she
had left after being
burned out. Age 8.

34

This small guest cabin, just uphill from the main house, was home to Mandy, Jessie, and Patty Trevorrow and Bob Haas through the summer of 1948.

Jessie and Patty Trevorrow with Bob Haas, in the middle of the new house construction. Looking over Patty's shoulder is Frank Haas, Bob's brother.

The new house was built in the pattern
of the old one. Patty wears a new dress
sent by the kids in Connecticut.

Bob Haas, shown here
in the summer of 1948, today
still spends hours hoeing.

(Bob

Bob and Pat Haas cut their wedding
cake, October 1958.

(*Brattleboro Reformer*, March 1996, Jim Powers)

Almost fifty years after the fire, Pat and Bob Haas attend the Westminster Town Meeting.

Michael and I leave the church after
our wedding, 1981. My mother looks
over Michael's shoulder, at the left of the
picture. My father stands behind me.

(Mary Amster)

chapter (*5*)

MOM HAS GONE BACK to the bedroom.

First she puts on her corset. Mom doesn't wear her corset to hold her tummy in. She wears it to hold her back together. She has worked on other people's farms and in other people's kitchens since she finished eighth grade. No one ever told her, "Jessie, be careful" or "That's too heavy for you to carry." She was there to do the heavy work. Now she has to put the corset on the minute she gets up, or her back will hurt so much she won't be able to move. Somehow, tonight, she ran downstairs without it. She's lucky she can still walk.

She puts the corset on, right over her nightgown. She puts a bathrobe over that. Then she throws her pocketbook onto the bed. Inside is all their money and her diamond ring. She sweeps up the pictures on the dresser, the combs and brushes, and all the clothes she can see, and throws them on the bed too. When she's ready to go, she will gather up everything in the blanket.

But first she wants to save her hope chest. Grandpa Yeaw made it out of wide pine boards, one board to each side. It holds everything Mom has saved for the house she and Dad hope to have someday.

The chest is heavy, and Mom knows she can't lift it to get it out the window. But maybe she can push it down the stairs in front of her.

"Jessie! Jessie!" Everyone downstairs is shouting at her. "Come down! Get out of the house!" The bedroom is at the end of the house nearest the fire. Mom doesn't know how quickly the fire is spreading, but the smoke is getting thick.

She tries to push the chest. It barely moves.

"Jessie! Get out! Come on!"

She gives the trunk another shove, but it's too big. She can't move it quickly enough. Even if she could get it over to the stairs and started down, it might get stuck. It might keep her from getting out of the burning house.

Everyone is shouting. They sound so worried that Mom gets worried, too. If she really is in danger, someone will come upstairs after her. Then that person will be in danger, too.

Mom grabs the nearest things, the radio and her knitting, and runs down the stairs. She leaves her pocketbook and everything else she meant to bring still piled on the bed.

Now Patty can see Mom again, and for a minute she feels better. If only somebody would come up the hill! If she had someone besides the teddy bear to hug! But everyone is still busy. There is still work to do.

Neighbors have been arriving. They drive up from East Dover village and down the hills from surrounding farms. They park their cars and run to help.

Al and Uncle and Bobby have dragged some furniture outside, but it's still too near the house. Priscilla has thrown a mountain of clothes out the bedroom windows, her own clothes and her parents'. Everyone grabs clothes and furniture, and they carry the things farther away from the house.

Dad brings out the long gilt mirror. It used to belong to his and Auntie's mother, who died many years ago when Dad was a little boy. Dad carries the mirror far down the road, out of everyone's way, and puts it carefully behind the stone wall.

As he hurries back, he listens for the fire truck. He can't hear it yet. It has a long way to come, and the dirt roads are deep with mud. Mud is cold on Dad's feet, cold all the way up his ankles, and full of little sharp stones. The heat from the house feels good to him.

In front of the house the lilac bush is burning, the beautiful lilac bush that was full of purple blossoms every spring. It is covered with tight, hard buds, but they will never open. The fire is killing them.

The blacksmith shop is ablaze, too, and little fires start all along the roofline of the henhouse. To Patty they look like footlights on a stage.

Mom runs into the burning henhouse. Two neighbors follow her. They push the hens out the door. They throw them through the open window. The chickens squawk and soar through the air with their wings flapping wildly.

The fires on the henhouse roof flare up quickly, big and bright. "Mom! Get out!" Patty screams. "Get out! Get out! Get out!"

They're done. They run outside and into the other chicken house, where the young hens live. It is farther from the fire, but the outside of it is smoking. Mom and the neighbors push the young hens out.

Then they come outside. They chase the sleepy hens across the lawn, toward the cool, dark edge.

Most of the house is on fire now, but Uncle thinks he could still save some of his guns. Everyone says *"No!"* It's crazy to go back into a burning house.

Uncle goes in anyway. After a few minutes he crawls back out on his hands and knees. The smoke was too thick, and the fire was too hot. Auntie and Priscilla are crying. They help Uncle to his feet and away from the heat.

Finally Mom comes up the hill. She comes slow and tired, and breathing hard, coughing. She hugs Patty, and hugs her again, and Patty hugs back, hard. She's still crying.

"Everything's all right," Mom says. "Nobody got hurt, and there's nothing to be afraid of."

chapter **(6)**

MOM LEANS AGAINST THE CAR and holds Patty against her. They watch the burning house.

Everyone else watches, too. There's nothing more to do. Bobby sits on the stone wall, and Rover sits close to him. Bobby puts his arm around Rover. The dog's body shakes. They can hear the fire trucks now, roaring and chugging up the muddy hill.

A sound begins within the burning house: a sigh, a groan, a long loud *whoosh*. The roof collapses into the cellar with a great crash. Sparks and flames shoot up toward the sky. The fire sounds like a laughing wind, hotter now, so that all the people have to move back even farther.

One end wall is left standing, rimmed with flame. The windows are still there. One of those windows was in Al's bedroom, but there is no bedroom left. The wall stands all alone, thin and bright with flame against the black sky.

Over the fire's roar comes the sound of gunshots. Everyone looks around. But the sounds come from inside the house. The rifle shells and shotgun shells are exploding in the heat.

At last the fire trucks pull into the yard, engines straining. The firemen start to unroll their hose. They have to run the hose a long way to get water, down a steep bank to the brook. They want to put water on the last wall of the house. Maybe they can save that.

But Uncle says no. He points to the barns and argues with the fire chief. Finally the firemen turn their hoses on the roofs and walls of the barns. Steam rises in big clouds. The barns were very hot, but now that the fire trucks have come, they will be saved.

"My pigs!" Dad says. "I forgot to let my pigs out!" But it doesn't matter. The barns won't burn. The pigs are safe.

But the washtub was left on the porch. No one saw the ducklings inside it, and they are dead.

The sky is getting light now. The fire seems smaller and paler, sunk down into the cellar hole. It has eaten everything that could carry it high into the air. Now it can only smolder and glow, heat the foundation stones, and make the shotgun shells explode.

There is nothing left to do. The firemen will watch the barns and the chicken house. They'll make sure the fire doesn't spread, and it will get smaller and smaller.

While Mom and Patty watch the fire, Mr. and Mrs. Rogers talk to Uncle and Auntie. Then everyone starts up the hill toward the cars. Auntie is still in her nightgown. Priscilla is in her night-gown, too, but she has her dungarees on under-neath. Uncle and Al wear their pajamas, and Bobby has pants and boots on. Dad limps behind them all on his bare, hurt feet.

"We'll go on down to Rogers's and have some breakfast," Uncle says. He sounds quiet and tired. Patty can see his eyebrows and his front hair, curled and frizzed from the heat of the fire.

Al and Bobby drive down in the farm truck.

Uncle drives the car, and Auntie and Priscilla ride up front with him. Mom, Dad, and Patty squeeze into the back.

Patty is happy to be squashed between Mom and Dad. They smell like smoke and they are dirty, but they are all right. Nothing terrible happened to them. Mom's back didn't even go out.

"I left my pocketbook on the bed," Mom says, "with my diamond ring and every last bit of our money!" Then she laughs. "But I saved my knitting!"

Patty snuggles against her. Mom gives her a hug. "You woke us all up, Patty," she says. "I guess you saved our lives."

"I guess she did!" says Dad.

"I just wanted a drink of water," Patty says.

But it is true. If she hadn't asked for a drink of water, Mom would have kept on sleeping. If

Mom hadn't gone to the window and seen the smoke . . . if Dad didn't need to sleep with the window open in the first place . . .

We *all* saved our lives, Patty thinks.

But they couldn't save the house. It is gone forever.

The Rogerses live in an old farmhouse down the road. Mrs. Rogers makes breakfast. She is the same size as Mom and Auntie, and she gives them dresses to wear.

She gives Patty and Priscilla dresses, too. Patty's dress is so huge she can hardly walk in it.

It's fun to sit down to breakfast in a different house, wearing strange clothes. The sun shines in through the windows, and the bacon smells good. Everyone is so glad to be alive, and they all have stories to tell about what they saw or what they did. All over again, around the kitchen table, Patty wakes up coughing, and Mom says, "Don't be afraid"; the pump handle comes off in Dad's hand; he tramples the wire with his bare feet; Mom runs downstairs with her knitting and the radio; up by the car Patty screams.

"I saw the piano burning," Al says. "The outside went first, and you could see the skeleton, all the wires."

Auntie looks around at Mrs. Rogers's clean kitchen. She thinks of her own kitchen, with the pantry and the broom closet. Just last night she shut the door on that kitchen. She will never open it again.

Priscilla keeps sniffing her hair. It stinks of smoke. If she could wash it, maybe she would feel a little better.

Uncle is quiet, too. After a while he starts to cry.

Patty has never seen a grown-up man cry before. It's almost as frightening as the fire. "Don't cry, Uncle," she says. "Please don't cry."

But Uncle thinks it is his fault the house is gone, because he put the pump together wrong. He has lost all his carpenter's tools, all his machinery. He has lost the house he grew up in, the house he had from his father Frank Yeaw, and *he* had from *his* father, Delphia Yeaw. Uncle has lost things he can't even remember right now. He has to cry.

Patty has lost all her clothes except her pajamas. She has lost the old china doll and her bank that looked like a Scotty dog. She has lost her books and her toys, her picture of a rabbit, and the matchbox houses where her imaginary fur people lived. She has lost her Lone Ranger ring that was in Dad's jar of half-dollars. She has lost the house that she grew up in, too.

But she has already cried, and besides, Mom and Dad are all right, Auntie and Uncle are all right, Priscilla and Al and Bobby are all right, Rover and Tommy Tiger are all right. Everything is terrible, but not as terrible as it might be, and it's also a little exciting.

Friends come up from the village with clothes and food. When Dad and Bobby have something to wear, they go up the hill to feed the animals. Chub and Sailor and the sheep come back to the barn. The older cows come, too. But the heifers won't come. The pasture gates were all open, and the young cows have escaped.

When they've finished chores, Dad and Bobby

stand for a while looking at the place where the house used to be. It is just a big heap of ash and coals. It's still very hot, and every once in a while a maple syrup can explodes.

The syrup was stored in the attic. But the attic of the house has fallen into the cellar hole, and all the heat of the fire is on top of it. Each time a can blows up, a geyser of maple syrup shoots up fifteen feet in the air and splashes and hisses on the hot coals. The air smells like burned maple candy.

Rover leans against Dad's leg and whines softly. Rover likes everything to be orderly and normal. He doesn't understand what has happened.

"I bought you a birthday present yesterday, Mandy," Bobby says to Dad. He points to the ashes. "It's in there somewhere."

"But we're not," Dad says.

Dad remembers the mirror now. He and Bobby

go to find it. They walk up and down the stone wall twice. But there is no mirror anywhere. Dad is upset. He was very careful with that mirror. He remembers right where he put it. How can it be gone?

But when he gets back to the Rogers's house and tells Auntie, she says, "It's all right. Somebody took it down to the village for safekeeping."

Al calls home to Glendale, to tell his family the news. His train ticket was burned up, with all of his money. But Bobby has money to loan him.

Auntie, Uncle, and Priscilla drive up to the house. They find the clothes they have saved, piled out on the lawn. They go into the small guest cabin just uphill from the house to change. Then they start to take the rest of their things across the road to the barns. Al and Bobby come up to help.

Dad goes out to look for the heifers. Mom lies down on Mrs. Rogers's couch. Patty sits beside her and hugs the bear. He smells like smoke, and so does Mom.

Mrs. Rogers is working in her kitchen. Patty could go in and help her and talk. But she wants to be with Mom, even though Mom is asleep.

No one else is around. That seems scary, almost as scary as the fire. The house is gone. Without the house, what will keep everyone together?

chapter (7)

WHAT HAPPENS NEXT? Does everything come to an end? No more farm? No more summer boarders?

No. Already that afternoon, Uncle begins to plan a new house. He had insurance. That will pay him some money. He is a carpenter, so he can do lots of the work himself. He can cut trees on his own land for some of the lumber. The new house will stand where the old one stood, and it will be ready to live in this summer. He'll start to work as soon as the ashes cool, and as soon as they have sifted through to find anything that didn't burn.

There is hardly anything to find. The head of the china doll. Dad's half-dollars, melted, and coated with melted glass.

While the new house is being built, Uncle, Auntie, and Priscilla stay with a neighbor. Dad and Mom, Patty, and Bobby spend the summer in the little guest cabin, near where Patty stood and watched the fire. Now Patty can watch the new house being built.

In the evening, after their farm work is done, Dad and Bobby go down to the new house to do more work. Mom and Patty go down, too. Patty likes to play with the leftover pieces of wood. Later, when the plumbing is hooked up, she can take a bath in the bathroom with no walls.

It's fun staying in the tiny cabin, with the living room and kitchen all in one room, and sheets hung up in the corners of the bedrooms to make closets. They joke and laugh together. Bobby builds huge sandwiches with a slice of everything in them. Once Dad puts in a banana skin when Bobby isn't looking, and Bobby eats his sandwich without noticing anything different about it.

It rains often that summer. They say to one another that they can't get to sleep without the sound of the rain on the cabin roof.

Two neighbors who teach school in Connecticut, Helen Hudson and Alice Nelson, help Uncle and Auntie outfit the new house. They find good furniture that people don't want anymore. A restaurant that is buying a new set of dishes gives

Auntie its old china. It's heavy white china with a pattern of green leaves, more dishes than Auntie could ever use. There is enough to give away, and still there are dishes left over. Bobby and two neighbors drive down to Connecticut in the farm truck and bring all these things back.

The two teachers tell their Connecticut classes about the little Vermont girl who lost all her clothes and toys in a fire. The girls at those schools go home and clean their closets. They send huge boxes up to East Dover, full of clothes, books, toys, a cooking set. Patty ends up having more things than she ever did before.

Many of the new clothes don't fit. Patty asks her friends from school to come up and try them on. All the kids in East Dover get new clothes that summer.

A neighbor gives Auntie some old featherbeds to use for making pillows. Mom sews new pillows out of soft cotton grain sacks. Then she and Auntie open the featherbeds out in front of the cabin. As they stuff the pillows, the feathers drift onto the bright green grass. To Patty it looks like snow in summer.

The boarders come, even before the new house is finished. By July the first ones arrive. They have to walk up slanted planks to get inside, because there are no steps yet. The farm isn't as restful that summer, but the boarders don't mind. They are friends by now, not just customers. They know Uncle and Auntie need the money they are paying. They want to see the new house being built, because for all the summers to come it will be their house, too.

When the new house is finished, it looks almost the same as the old house did. It is a little bigger and a little easier to work in. It is not the same. It is not as beautiful. It is not the house that Uncle and Dad and Patty grew up in. But it is a good house.

The well beneath the house is still there, but the water is yellow now, full of ashes and burned paint. It never clears. It's only good for keeping things cold. Dad keeps the cream cans there, but no one can drink the water that used to taste so good. Uncle digs out a new spring, but it isn't the same.

The heifers go out to the summer pasture, on the hillside with the bobcats and deer. In the fall Dad can't catch them. They were spooked the night of the fire and ran loose for a long time afterward. Dad didn't get a chance to gentle them. They will never be good milk cows now, and they can't live through the winter in the wild. Dad has to shoot them.

In a way it is the fire that kills them, and it makes Dad and Patty sad. But they can eat the meat. There is a new house to live in that winter, and a new kitchen, and everyone is there to sit and eat roast beef. Out in the barn, new calves grow inside their mothers, and in the spring they will be born.

The lilac is gone, and everyone misses it. But a few years later it suddenly comes to life. It sends up new shoots out of the ground, and Patty puts a fence around them, so no one mows them off. She tends the lilac, and it grows full and tall again, and blossoms in front of the new house just as it blossomed in front of the old one.

afterword

In 1958, ten years after he carried her out of the burning house, Bob Haas married Pat Trevor-row. They bought a farm of their own, in West-minster, and in their new living room they hung the long gilt mirror that Patty's father had hidden behind the stone wall.

In 1959 I was born, and named Katherine

Jessie, after my two grandmothers. Two years later my brother, Jim, was born, and four years after that my sister, Martha.

We were brought up on stories about the farm in East Dover—stories about Chub and Sailor and Rover, about bears and bobcats.

But the big story was The Fire. Everything dated from that. If we asked our mother about something that happened when she was very small, and she didn't remember clearly, her eyes would widen and she would say, "But that was back Before The Fire." *After* The Fire was modern life, and the beginning of growing up.

Jim and I heard about The Fire when we stayed with our grandparents in East Dover, too. Early in the morning Gram would send Gramp off to work at the ski area with his two sandwiches, one marshmallow fluff and one dried beef. Then we would pile into bed with her, watch the world outside get light, and beg for stories. First we wanted the bear stories, and then we wanted to hear about the fire. Later, when we got up, we could play with the melted lump of half-dollars. When we got home, we could look at the limp old teddy bear in our mother's closet.

That fire changed more than my family's life. Because of it, farmers all around southern Vermont built fire ponds. In wet places near the farmhouses they bulldozed big holes, which filled with water. People swam in the ponds, cattails grew, frogs and ducks came to live. And if a fire started on one of those farms, many times it was put out quickly, because the water was there.

Because it took so long for fire trucks to come from Brattleboro, the people in East Dover started a fire department of their own. They held bake sales and fairs to raise money. They built a fire-house and bought trucks, and men volunteered to be firefighters. My grandfather volunteered. He and my grandmother bought a house of their own, down the road from Uncle's farm. A sign outside the new house said *Deputy Fire Warden, Mandy Trevorrow*.

My parents also became volunteer firefighters in Westminster. My mother was one of the first women on the force, and later she was Deputy Fire Warden, just like her father.

Every year my mother went into the school and told the children what to do in case of a fire. Don't *ever* go back inside a burning house, she told them. There's nothing in a house that's worth risking your life for. My mother knows. She saw her mother and father go back into a burning house. She waited for them to come out.

I grew up and went to Wellesley College, and I started writing stories. One afternoon I went to lunch with my mother, a neighbor, and two teachers from Wellesley. As we sat around the table, my mother began telling them the story of The Fire. When she got to the part where her mother tells her, "Don't be afraid. The house is afire," tears came to my eyes. I thought, I'm going to write *that* story. And I did.